I0188637

UNIFORMS OF THE SAXON ARMY 1699-1764

ILLUSTRATED BY
CAPTAIN BOMMER

TEXT BY
G.F.NAFZIGER

UNIFORMS OF THE SAXON ARMY 1699-1764

ILLUSTRATED BY
CAPTAIN BOMMER

TEXT BY
G.F.NAFZIGER

THE NAFZIGER COLLECTION

Uniforms of the Saxon Army 1699-1764
By Captain Bommer
Translated by George Nafziger
This edition published in 2023

Winged Hussar Publishing is an imprint of

Winged Hussar Publishing, LLC
1525 Hulse Rd, Unit 1
Point Pleasant, NJ 08742

Copyright © Winged Hussar Publishing
ISBN 978-1-958872-16-1
LCN 2023934401

Bibliographical References and Index
1. History. 2. Germany. 3. 18th Century

Winged Hussar Publishing, LLC All rights reserved
For more information
visit us at www.whpsupplyroom.com

Twitter: WingHusPubLLC
Facebook: Winged Hussar Publishing LLC
Instagram: Winged Hussar Publishing LLC

This book is sold subject to the condition that it shall not, by way of trade or otherwise, be lent, resold, hired out, or otherwise circulated without the publisher's prior consent in any form of binding or cover other than that in which it is published and without a similar condition, including this condition, being imposed on the subsequent purchaser.

The scanning, uploading, and distribution of this book via the Internet or via any other means without the permission of the publisher is illegal and punishable by law. Please purchase only authorized electronic editions, and do not participate in or encourage electronic piracy of copyrighted materials. Your support of the author's and publisher's rights is appreciated. Karma, it's everywhere.

INTRODUCTION

As I was searching through the *Bibliotheque nationale's* website I ran across seven folios of watercolors of the Saxon Army's uniforms from 1699 to 1764 by a Captain Bommer who went through the "Military Archives of the Kingdom of Saxony" and painted watercolors from what he found there. These folios were not dated, but apparently were done before 1870, when the Kingdom of Saxony was absorbed into the newly forming Germany.

There was no other explanation of these works. Though numbered they were only in a vaguely chronological order and the order in which they were organized has been maintained.

Bommer gave no description of the Saxon Army or its organization. That information has been added to provide a better understanding of the Saxon Army. The principal sources of this data were:

Anonymous, *Geschichte des Königlich Sächsischen 6. Infanterie-Regiment und seine Vorgschichte 1701-1887* (Leipzig: Commissions-Ver Giesede & Devrient, 1887).

Anonymous, *Geschichte des Königlich Sächsischen Königs-Husaren Regiment No. 18* (Leipzig: Baumert & Ronge, 1901).

Hauthal, F., *Sächisischen Armee in Word und Bild* (Leipzig, J.G. Bach, 1859).

Knotel, R, Knotel, H. and Sieg, H. *Uniforms of the World* (New York: Scribner, 1937).

Müller, Richard. *Die Armee Augusts des Starken: Das Sächsische Herr von 1730 bis 1733* (Berlin: Miltärverlag Der Deutschen Demokratischen Republik, 1984).

Schönberg, G., *Geschichte des Königlich Sächsischen 7. Infanterie-Regiments „Prinz Georg" Nr. 16* (Leipzig: Brockhaus, 1890).

Schuster, O. and Francke, F.A., *Geschichte der Sächsischen Armee* (Leipzig, Ducker & Humbolt, 1885).

Wagner, *Geschichte des Königlich Sächsischen 8. Infanterie-Regiment „Prinz Johann Georg" Nr. 107 1867-1891* (Leipzig: Dürr, 1893).

CONTENTS

THE SAXON ARMY
1699-1764

The Saxon Army had undergone a consistent growth since the Feudal era and standing regiments and military formations started to appear. In the 1670s the Elector of Saxony continued this process and increased his guns and artillery, strengthened the fortifications of several towns and castles and constructed new ones, in order to be prepared to deal with any contingencies in an unstable Europe.

George III, who gained a great deal of war experience in his father's campaigns and recognized the great value of a war constitution based on wise and firm laws, put the Saxon Army of 12,659 men on a regular footing, abolished the Croatian bodyguards, which had become superfluous, and set up several standing cavalry and infantry regiments. In 1683 the Elector sent 12 regiments, which consisted of half cavalry and half infantry, to assist in the breaking of the siege of the Vienna, which was besieged by the Turks, where they gained a share in the victories won over the Ottomans. After the end of this campaign, the Saxon troops returned to their homeland, but they were already moving 1686 back to Hungary and 1688 to the Rhine, to fight enemies of the German Reich.

Under the brief reign of the Elector John George IV, the Saxon Army was again increased, not just with a few regiments, but also with a cadet company and the "Grands-Mousquetaires"[1], which his brother Augustus II, immediately merged with the Klemm and Schöning Dragoon Regiments after he came to power. This was followed by the most important period for the Saxon armies, their existence and their formation, which was characterized by the most varied and frequent changes. The Kings of Poland, compelled by repeated outbreaks of unrest and wars, had to think about a significant increase in their armed forces just as often, when, after the suppression and end of that unrest, the need of the ruined and devastated country, where economics had compelled the partial dismissal of their. Only by means of a large army, which became even stronger through good organization and discipline, could Augustus II succeed in resisting the enemies who appeared on the frontiers of his new country, particularly King Charles XII of Sweden. It was therefore necessary to raise several new regiments, especially since a not insignificant force, 12,000 Saxon troops, had been with the Imperial army in Hungary since 1695, and was not withdrawn until 1699. It was fortunate for Saxony that the increase in its army in 1697 produced a better military and economic establishment, which included the establishment of a General Staff. In 1700 there were only 41 squadrons of cavalry (including 18 squadrons of dragoons) and 24 battalions of infantry under the supreme command of Count Flemming in Lievland when Riga was blockaded, by 1703 the army had already grown so much that it had 8 regiments of the Guard, 8 regiments of cuirassiers (32 squadrons), as many regiments of dragoons, 12 regiments of infantry (24 battalions), 6 regiments of home defense forces, about 5000 men, 2 regiments of knightly horse, each of 900, and 1 battalion of artillery. The defense forces were replaced by the land militia in 1711, and these regiments were dissolved altogether in 1716.

By the year 1706, six new regiments, four mounted and two on foot, had been raised; but after the Peace of Alt Ranstädt there was a great reduction in troops, as their presence in Poland was now superfluous, and 9,000 men were sent to the Lower Rhine, where 6,000 Saxon troops in Dutch pay were already stationed. These troops remained in the Netherlands until the Peace of Utrecht and earned praise, especially the Chevalier Guard, which consisted of 80 men, who fought bravely in the battle of Frauenstadt, which they lost despite the excellent orders of Schulenburg, and recaptured their silver drums.

---

[1]There are a few possible translations of this: Grand Musketeers, Great Musketeers, and Giant Musketeers.

After the defeat of Charles XII and when at Poltava the superiority of the Swedish party in Poland had sunk, Augustus II seized the Polish Crown again, invading the country with 11,000 cavalry. As this war had constantly demanded more forces, new regiments had to be raised because of lack of troops in the Netherlands. Gradually four regiments of infantry and a regiment of dragoons were formed, and the Chevalier Guard was reduced to four brigades, each of 60 men. For the possible protection of the hereditary lands against the intrusion of enemy troops, apart from the knight horse, there began a levy of men aged 20 to 40 years, thus in 1710 Saxony had an army of more than 54,000 men. However, many men were dismissed from active service after only 4 months, since the fear of a Swedish invasion soon proved to be unfounded.

Some bloody battles were followed by the pacification of Poland by the conclusion of the peace with Sweden, which made it possible for the Saxon troops to return to their homeland and this entailed their immediate release from service.

In September 1714 the infantry regiments consisted of:

	Staff		Company
1	Oberst	1	Captain
1	Oberstleutnant	1	Leutenant
1	Major	1	Fähnrich
1	Regimental Quartermaster	2	Sergeants
1	Adjutant	1	Quartermaster
1	Regimental Judge	1	Surgeon
1	Regimental Surgeon	5	Corporals
4	Gefreite-Korporals	3	Drummers
5	Musicians	75	Gefreite and Privates
1	Provost with servant		
18		90	

The Order of June 20, 1717, ordered the disbanding of several cavalry and infantry regiments. When they were finished the Saxon Army contained:

Guard: Chevaliers-Garde, Garde du corps
Cavalry:
    Cuirassiers Königlicher Prinz, Prinz Alexander, Pflugk, and Criegern
    Dragoons: Baudissin, Unruh, Bielcke, Birkholz, Klingenberg
    Hussars: 1 regiment
Infantry:
    1st Garde, 2nd Garde, Königlicher Prinz, Weissenfels, Diemar, Fietzner,
    Pflugk, Drossky, and Marschall
Artillery:
    Hausartillerie, Field Artillery, and Artillery Battalion
Special Troops:
    1 company of pontoniers, 1 company of miners

This reduction took place in such a way that four companies were dismissed from all infantry regiments, which reduced the number of each individual regiment to eight companies of 91 men each. It was similar with the cuirassier and dragoon regiments, where each regiment lost two companies, leaving them with only six companies that contained 80 privates. The Königin, Leib Regiment, Wolfersdorf, and Count Moritz von Sachsen Cuirassier Regiments, were completely abolished with some men being dismissed and others being assigned to other troop detachments. The same fate was suffered by the Leibdragoner, Bayreuth, Brause and Saintpaul Dragoons, the Seydlitz Infantry Regiment together with the 4 companies of the Maier Freidragoner, finally the Haiduck company, the 3 companies of the Wittenberg garrison and the Dresden garrison regiment. The two Flemming Infantry Regiments were incorporated into the Polish Crown Guard and the Miers Dragoon Regiment was left to the Republic of Poland as a future Royal Guard. The troops of foreign sovereigns who were in Saxon pay were dismissed and returned.

As a result of all these reductions, the size of the Saxon Army fell back to two Guard regiments, 4 cuirassier regiments, 5 dragoons and 10 infantry regiments, of which 6,000 men went to Hungary in 1718 under the supreme command of Duke Johann Adolph von Sachsen-Weissenfels.

A decree dated June 20, 1717 set the organization of an infantry company at:

Staff		8 Companies	
1	Oberst	6	Captains
1	Oberstleutnant	8	Lieutenants
1	Major	8	Fähnrichs
1	Regimental Quartermaster	24	Sergeants
1	Adjutant	8	Quartermasters
1	Regimental Surgeon	8	Surgeons
5	Musicians	48	Corporals
1	Provost	16	Drummers
1	Provost's servant	80	Grenadiers
		48	Gefreite-Corporals
		8	Fifers
		8	Sappers
		504	Musketeers
13		**726**	

Now there was a longer break in Saxony's involvement in wars: the country, the bourgeoisie and the peasantry, recovered in 15 years of calm. The hands, which had been withdrawn from industry and agriculture by the ongoing war were returned to the fields and workshops as a result of the peace, turned to business with renewed activity, and the increasing prosperity of the country. The subsequent increase in Saxony's inhabitants prompted the King of Poland [also the Elector of Saxony] to establish a Grands-Mousquetaires Squadron, a Leibgrenadier Garde, a Janissary and a new grenadier regiment. He also increased each cuirassier regiment by 159 men, each dragoon regiment by 107 men, and each infantry regiment by 706 men, bringing the effective strength of the Saxon military force back to 7,047 horse and 19,415 infantry, i.e. a total of 26,462 men.

After the death of Augustus II ("The Strong") in 1733, his son Augustus III, who was also the elected King of Poland, led the army, reinforced by two regiments of light horse [Chevauxlegers], into that country to quell the new disturbances and to hold down the old factions who continually fought each other, while four newly formed regiments of land militia, each 1,979 men strong and each consisting of 4 grenadier and 8 musketeer companies remained to protect the Saxon hereditary lands.

From 1733 to 1745 it was indeed found expedient to reduce the strength of dragoon regiments to 490 men and that of the cuirassier regiments to 484 men and to restructure them into two squadrons, each of three companies; but at the same time two regiments of light horse [Chevauxleger], also called a "Korps der Jäger zu Pferde" and various infantry regiments were raised and the existing ones were increased by the addition of 255 men, namely with two grenadier and four musketeer companies, so that the grenadiers, which were converted into grenadier battalions in the event of a war. This meant that purely grenadier companies were formed, and the grenadiers were no longer distributed among all infantry companies as before.

Each of the Chevauxleger Regiments contained:

	Staff		Per Company
1	Oberst-Leutenant	1	Captain
1	Major	2	Lieutenants
1	Senior Regimental Judge	1	Fähnrich
1	Adjutant	2	Wachmeisters
1	Provost	1	Fourier
1	Servant	1	Surgeon
		10	Corporals
		1	Saddle maker
		1	Blacksmith
		2	Drummers
		132	Cavaliers
6		154	

On May 15, 1735 this structure was changed as follows:

	Staff		Per Company
1	Oberst	1	Captain
1	Oberst-lieutenant	1	Staff Captain
1	Major	1	Lieutenant
1	Senior Quartermaster	1	Fähnrich
1	Adjutant	2	Wachmeisters
1	Senior Regimental Judge	1	Standard bearers
1	Regimental Surgeon	1	Fourier
1	Provost with servant	10	Corporals
1	Provost Servant	1	Saddle maker
		2	Drummers
		1	Blacksmith
		131	Cavaliers
		1	Provost Servant
9		154	

The 1st and 2nd Companies were designated as Schützen Companies and the 3rd and 4th Companies were designated as Dragoon Companies.

A new state for the light cavalry, now called light dragoons, was established and consisted of:

	Staff		Each Company had:
1	Oberst	1	Captain
1	Oberst-lieutenant	1	Premier-lieutenant
1	Major	1	Sous-lieutenant
1	Adjutant	1	Fähnrich
1	Regimental Quartermaster	2	Wachmeisters [Sgt. Majors]
1	Judge	1	Standard Bearer
1	Regimental Surgeon	1	Fourier
1	Baggage Master	1	Surgeon
1	Drummer	10	Corporals
8	Musicians	2	Drummers
1	Provost with servant	1	Saddle makers
		1	Blacksmith
		131	Men
		1	Provost Servant
18		155	Total

Augustus III, whose troops had suffered substantial losses in the 1745 campaign resulted in the army's overall strength being greatly reduced. This was addressed by gradually raising entire regiments, such as the L'Annonciade Cuirassier Regiment in 1746, the Minkwitz, O'Byrn, Graf Ronnow and Dallwitz Cuirassier Regiments in 1748; also the Prinz Sondershausen Dragoon Regiment, the 2nd Garde Regiment and the Jasmund, Allapek and Bellegarde Infantry Regiments. These were assigned by companies to the other regiments, so that the cavalry gained a strength of 12 and the infantry 18 companies, while the Leib Companies of the reduced regiments were entirely dissolved.

In 1740 each infantry battalion raised a grenadier company that consisted of:

1	Hauptmann (Captain)
1	Premier lieutenant
2	Sous lieutenant
3	Sergeants
1	Quartermaster
6	Corporals
2	Fifers
2	Drummers
6	Sappers
1	Surgeon
96	Grenadiers

After the Second Silesian War (1744-1745), the Saxon Army had undergone severe reductions. On November 25, 1747, five cavalry and four infantry regiments were disbanded, and their manpower was used to rebuilt the other regiments. The infantry regiments disbanded were the 2nd Garde, Jasmund, Alluspeck, and Bellegarde Infantry Regiments. The 8 grenadier companies were to be combined into a grenadier battalion under Major General von Bellegarde; A company was to be formed from the semi-invalids to guard the prison at Waldheim. Each of the 11 remaining regiments received an addition of 4 companies. The regiments now consisted of 2 grenadier and 16 musketeer companies. The organization of these grenadier companies was:

1	Captain
1	Premier-lieutenant
2	Sous-lieutenant
2	Sergeants
1	Quartermaster
1	Surgeon
5	Corporals
2	Drummers
2	Fifers
2	Sappers
<u>72</u>	Grenadiers
91	Total

Each musketeer company contained:

1	Captain
1	Premier-lieutenant
1	Sous-lieutenant
1	Fähnrich
2	Sergeants
1	Gefreite-Corporal
1	Surgeon
5	Corporals
2	Drummers
1	Sapper
<u>72</u>	Musketeers
89	Total

At the beginning of 1748, four more infantry regiments were formed, their companies were divided into the others, but with reduced manpower. Among these the Fankenburg Regiment was assigned four musketeer companies from the disbanded Bellegarde Regiment.

The infantry then consisted of 20,128 men in a grenadier battalion and 12 regiments each with 18 companies (2 grenadier companies of 91 men each and 16 musketeer companies of 89 men including officers).

The army consisted of 30,336 men, excluding 2,518 uhlan (or Tartars) and 7,920 circle troops.

On January 1, 1749, a new budget was established for the Army. In order to meet it the commission ordered that the Carabinier-Gardes and the six cuirassier regiments be reformed with eight companies of 40 privates, while the 11 infantry regiments were to form 12 companies of 96 privates. The grenadier battalions were reduced to 5 companies of 96 privates.

The artillery battalion retained 4 companies but, like the house company, they were reduced by 10 senior gunners and 20 junior gunners. In the Leibgrenadier Garde, 2 companies were to be completely disbanded and in every company of the cavalry 10 reiters were to be dismounted.

In general, in the infantry companies the junior captains were to be dismissed.

Each infantry regiment was reduced by the following:

3 Captains
3 Capitan-lieutenants
6 Premier-lieutenants
6 Sous-lieutenants
6 Fähnrich
6 Fahnenjunker
6 Quartermasters
6 Surgeons
2 Assistant Surgeons
18 Corporals
6 Sappers
<u>144</u> Privates

Total: 211   men

This process reduced the Army's infantry by 2,721 men, the artillery by 201 men, and the cavalry by 400 men and 288 horses.

In 1750 there was another reduction, and the cavalry was reduced by 80 NCOs and 800 men, the infantry was reduced by 132 NCOs and 2,670 men, the grenadier regiments by 5 NCOs and 50 men, and the artillery by 5 NCOS and 50 men. This meant that each infantry company was reduced by 1 sergeant and 20 privates.

By 1753, a Saxon infantry regiment consisted of: 55 officers, 118 NCO, 8 musicians, 36 drummers, 4 fifers, 14 sappers, 152 grenadiers, 700 musketeers, 12 surgeons, a junior staff of 6 men for a total of 1,105 men.

The regimental staff consisted of 17 men. Each regiment had 2 battalions of 5 companies (95 men each) and a grenadier company (97 men).

There were two exceptions to this. The Kurprinz Grenadierbataillon had only 5 companies for a total of 539 men; while the Leibgrenadiergarde had 14 companies organized into 2 battalions for a total of 1,684 men.

n 1754, the Saxon Army was reorganized into two Generalate under the command of Field-Marshal Duke von Sachsen-Weissenfels. He was succeeded at the head of the army by Count Brühl who immediately reduced the size of the army.

In 1755 instructions had been given to reduce the companies of cavalry to 30, and of infantry to 49 men. However, the increasing uncertainty of the political situation, which preceded the year 1756, made itself felt in such a way that the existing arrangement was approved again.

By 1756, the Saxon infantry was now mostly composed of Saxon soldiers and of soldiers recruited in countries neighboring Saxony.

At the outbreak of the Seven Years' War, the Saxon Army counted 12 infantry regiments with a total of 24 battalions and 1 standing grenadier battalion. There was also a peacetime establishment of 4 regiments for the Reichsarmee (only 180 men each).

In addition, there were garrison companies in Wittenberg, Königstein, Sonnenstein, Stolpen and Pleissenburg and Invalids in Waldheim. Altogether, these troops formed 8 companies for a total of 1,150 men.

For campaigns, the grenadiers of each regiment were combined into 6 converged grenadier battalions who, together with the standing grenadier battalion were organized into two brigades.

The grenadiers of the army were amalgamated into 7 battalions including Kurprinzessin. The 1756 wartime brigading of the grenadier battalions for the Pirna campaign was:

1st Bennigsen (2 coys Garde & 2 coys Graf Brühl)
2nd Kavannaugh (2 coys Prinz Friedrich August & 2 coys Lubomirski)
3rd Pforte (2 coys Prinz Xavier & 2 coys Gotha)
4th Götze (2 coys Prinz Maximilian & 2 coys Minkwitz)
5th Milkau (2 coys Königin & 2 coys Rochow)
6th Pfundheller (2 coys Prinz Clemenz & 2 companies of flank grenadiers of the
        Leibgrenadiergarde)
7th Kurprinzessin (5 coys Kurprinzessin)

The 1756 reduction of the army, of which only the Garde du corps, the four regiments of light horse [Chevauxleger] and the Leibgrenadier Garde were excluded, partly due to the reduction of the cavalry regiments from 12 to 8 (32 squadrons) and the infantry regiments from 18 to 12 companies, partly due to the regular annual loss of troops due to their end of obligatory service, who never returned to the flag, increased to such a degree that in the fateful year 1756, an infantry regiment consisted of only 50 officers, 120 NCOs, 4 fifers, 38 drummers and 846 privates, which were divided into 10 musketeer and 2 grenadier companies, each of 73 men.

The rest of the army consisted of 5 companies of artillery, 8 companies of garrison troops, and the smallish cadres of 4 Kreisregimenter (provincial militia) for a total of some 21,200 men. Furthermore, 4 cavalry regiments (Karabiniersgarde and 3 regiments of Chevauxlegers) with some 2,300 men and 2 Pulks (bands) of Tartar *Hoffahnen* (court-banners) with some 876 men were stationed in Poland in 1756, and thus, avoided the poor fate of their brothers-in-arms when the entire army surrendered at Pirna on October 15th.

When the army was assembled at Struppen, it consisted of 3,665 cavalry, 14,599 infantry, and 608 artillerymen. In addition, there were 155 cadets, 75 engineers, miners and pontoniers and 645 men from the garrison companies of Wittenberg, Waldheim and Pleissenburg; finally, the garrisons from Königstein, 195 men, and from Sonnenstein, 125 men; all together 20,066 men.

The artillery consisted of 4 26-pdr howitzers, 12 24-pounders, 27 12-pdrs, 4 6-pdrs and 50 regimental guns.

When the Seven Years' War began the Prussians quickly overran o Saxony and after the siege of Pirna, the Saxon Army surrendered in October 1756, and was forcibly incorporated into the Prussian army. From the Saxon troops that went into captivity Frederick II of Prussia raised Prussian regiments, but their morale was low, and they took every opportunity to desert. These deserters were gathered up and formed into complete units, at first in Bohemia and later in Upper Austria. So soon a corps of 10,000 men had gathered from such troops returning home, which participated in the campaigns in Hesse, Westphalia and on the Rhine in French pay, while the Carabiniers who had stayed behind in Poland, three regiments of light horse [Chevauxleger] and a few Uhlan Pulks in Bohemia and Silesia fought under the Imperial flag with many distinctions.

After the conclusion of the Hubertsburg Peace, which put an end to the Seven Years' War, these troops turned back to their homeland, and there was then a thorough reorganization of the entire army. The Elector assigned this task to Prince Xavier of Saxony, who had administered the Kurland during the minority of the later Elector Friedrich August, and he was not only concerned with increasing the army, but also with its good establishment and the procurement of all field needs. He used the experiences and advances which had been gained in the most important branches of the art of war by the various peoples of Europe since the beginning of this century. In connection with these efforts, which were accompanied by a good success, was the establishment of the artillery company in 1766, the renewal of the Order of St. Heinrich in 1768, and improvements in the basic instruction of the soldiers, which was particularly aimed at simplifying the indispensable hand movements. The Elector then had new drill regulations produced for the whole army, which underwent some changes but served as a guide for a long time.

At this time, the Saxon Army consisted of the following regiments: I. Cavalry: Garde du corps, Carabiniers, Electoral Cuirassier, v. Arnim, Prince von Anhalt, Count Ronnow, v. Penckendorf, v. Brenkenhof, the Feldjäger Corps, the Duke of Courland, Prince Albert, Count Renard, Baron v. Sacken Chevauxleger Regiment. II. Infantry: Noble Cadet Corps, the Swiss Guard, Engineer Corps, Artillery Corps, Leibgrenadier Garde Regiment, Kurfürst, Kurfürstin, Prinz Karl, Prinz Maximilian, Prinz Anton, Prinz Xavier, Prince Clemens, Prinz Gotha, Count v. Solms, v. Borcke, v. Thiele, and v. Block Infantry Regiment, the Semi-Invalid Company.

The Infantry

In 1683 the clothing of the Saxon infantry consisted of a coat with baize lining and either tin or brass buttons, a hat, cloth stockings of the color of the coat lining, and buckskin breeches. The Leib Regiment wore a red coat while the others wore gray. The coat linings were of various colors. Elector Johann Georg III had ordered that the infantry's pikes be placed in storage and that all the infantry be equipped with muskets and "swan's feathers", which could be used as musket rests. In 1686 the existing grenadier companies were replaced by a grenadier company that was attached to each foot regiment. The grenadiers then adopted blue, cloth caps. In 1687 the separation of the men into musketeers and pikemen was ended and red uniforms were issued in 1695.

The 10 infantry regiments that existed in 1700 were raised to 24 in 1701. Each regiment was formed with 13 companies including 1 grenadier company, the companies initially contained 72 corporals and privates and were soon raised to 120 men.

In 1701 the regimental distinctions were as follows:

Regiment	Distinctions[2]
Polish Guards	White
Saxon Guards	White
Königin	Isabella (cream)
Egidy	Isabella (cream)
Kurprinz	Lemon yellow
Thielau	Lemon yellow
Steinau	Green
Zeitz	Green
Bron	Unknown
Tromp	Unknown
Pistoris	Pale blue
Reuss	Pale blue
Sacken	Moss color
Marschall	Moss color
Fürstenberg	Dark blue
Löwenhaupt	Dark blue
Görtz	Sea green
Rothenburg	Sea green
Beichlingen	Gray
Weimar	Gray
Dönhof	Unknown
Flemming	Unknown

[2]The regimental distinctives come from Schuster and Franke.

In the Great Northern War, which started in 1702, the Saxon Army contained 27,000 men formed into:

General staff
Gardes du corps   4 squadrons 8 companies
6 Cuirassier Regiments   6 squadrons, 12 companies
4 Dragoon Regiments   6 squadrons, 12 companies
14 Infantry Regiments   13 companies each plus a grenadier company
Field artillery of 633 men.

In 1715 the Saxon officers adopted the gorget which was decorated with a coat of arms in yellow metal, and silver and crimson woven sashes, which were worn over the right shoulder. The coat had no lapels and was buttoned down the entire front so that the waistcoat was no longer visible. That same year the Janissary Corps was raised. Its uniform was a lemon-yellow coat, red waistcoat, and Hungarian-style breeches that were decorated with blue and white lace. They also wore yellow half boots and a Janissary cap (a turban for the officers). Their undress uniform consisted of green coats with yellow undergarments.

In 1730 the infantry had red coats with variously colored lapels and turnbacks, Swedish cuffs, and waistcoats. Their breeches were buff in color, their stockiness were white. A lace-edged halt was worn with a colored tuft. Their cartridge box was carried on a buff-colored shoulder belt. It bore no device for the musketeers, but the grenadiers had the Saxon coat of arms on their pouches and grenades in the corners. They also wore a match case on their belts and cartridge boxes on their waistbelts. They carried straight-bladed swords. The grenadiers wore a cap with a red front with a brass plate and a colored bag on the rear. In 1730 the distinctions were as follows:

Regiment	Coat Color	Distinctions	Buttons
Leibgrenadier Garde	Lemon yellow	Red	White
1st & 2nd Guards	Straw	Red	White
Kurprinz	Red	Lemon yellow	White
Weissenfels	Red	Yellow	White
Marchen	Red	White	Yellow
Löwendahl	Red	Pale blue	Yellow
Wilcke	Red	Cinnamon brown	Yellow
Saxe-Gotha	Red	Dark blue	Yellow
Böhl	Red	Straw	White
Caila	Red	Popinjay green	White
Weimar	Red	Green	White
Grenadier Company	Straw	Red	White

Four Kreisregimenter (district regiments) were raised in 1733 and they served until 1756, when they were disbanded. They wore red uniforms with blue distinctives. In 1734 the infantry adopted white coats; the Leibgrenadier Garde alone retained red ones. In 1743 the lapels on the coats of the infantry were abandoned and the coats became double-breasted. They had two rows of six buttons. The cuffs became plain and round. In 1745 colored collars were added to the coats of the officers and NCOs. The Kreisregimenter adopted gray coats at this time and all regiments had white breeches.

In 1745 the regimental distinctions were as follows:

Regiment	Coat	Distinctions	Buttons
Leibgrenadier Garde	Light red	Yellow	White
Foot Guards	White	Red	Yellow
Königin	White	Cochineal	Yellow
Kurprinzessin	White	Pale blue	Yellow
Friedrich August	White	Yellow	Yellow
Xavier	White	Pale blue	Yellow
Clemenz	White	French blue	Yellow
Brühl	White	Red	Yellow
Lubomirski	White	Yellow	Yellow
Rochow	Green	Red	Yellow
Minkwitz	White	French blue	White
Gotha	White	Pale blue	White
Friesen	White	Green	Yellow
1st Kreisregiment	Light gray	Yellow	White
2nd Kreisregiment	Light gray	Pale blue	White
3rd Kreisregiment	Light gray	Red	White
4th Kreisregiment	Light gray	Green	White

The Cuirassiers

In 1695 the basic color of the Saxon cavalry was red. Their cloak was of the same color. They wore tricorn hats. Their breeches were of yellow leather and they wore long boots.

Regiment	Distinctions
Leibregiment	White
Königin	Straw
Kurprinz	Yellow
Prinz Alexander	Green
Beust	Black
Eichstädt	Coffee brown
Damitz	Pale blue

In 1734 the Saxon cavalry adopted white coats. Their cuirasses were worn under their coats. The housings were of the distinctive color. In 1740 straw colored jackets and waistcoats were adopted. The NCOs now wore laced hats. The coats became double-breasted with two rows of eight buttons in each were adopted in 1741.

In 1754 the distinctions for the cuirassiers were changed to the following:

Regiment	Coat	Distinctions	Buttons
Garde du corps	Red	Pale blue	Yellow
Leibkürassiere	White	Bright red	Yellow
Kurprinz Cuirassiers	White	Pale blue	White
Arnim Cuirassiers	White	Crimson	White
Prinz Anhalt Cuirassiers	White	Yellow	White
Plötz Cuirassiers	White	Green	Yellow
Vitzthum Cuirassiers	White	Dark blue	Yellow

When the Saxon infantry was taken prisoner after Pirna, the cuirassiers formed themselves into grenadier companies that were taken into French pay. In 1765 new uniforms were adopted. The Gardes du corps received yellowish colored jackets and breeches, with blue collars, cuffs, skirt turnbacks, and waistcoats. They were all edged with yellow lace with red woven through it. Their cravats were red. Their undress uniforms and that of the officers' levee dress was red.

Dragoons and Chevauxlegers

In 1695 the Saxon dragoons adopted red coats, yellow leather breeches, and hats. Their distinctions, in 1707, were as follows:

> Bayreuth – Light blue
> Brause – Yellow
> Schulenburg – Straw
> Dünwald – Green
> Goltz – Black
> Wrangel – Coffee brown

Around 1730 grenadiers were added to the dragoon regiments. They wore caps like the Saxon infantry grenadiers. Their coats now had collars and cuffs. Their cravats were black, and their waistcoats and breeches were buff. At this time the distinctions changed again and were:

> Grenadiers à cheval – straw
> Arnstädt – Dark blue
> Katte – Poppinjay green
> Goldacker – Grass green
> Chevalier de Saxe - Pale blue.

The buttons on all these uniforms were white.

The Mier Dragoon Regiment wore polish dress. Its uniform changed slightly before the Seven Years' War. In 1754 the Rutowski Regiment of Chevauxlegers wore red coats with black distinctions, straw-colored waistcoats and yellow buttons. In 1765 the Chevauxlegers adopted red coats in place of their formerly green ones. The Albrecht Chevauxlegers had green distinctions and the Renard Chevauxlegers wore blue. They had a yellow metal buttons and straw-colored waistcoats and breeches. The Kurland Chevauxlegers, which were initially clothed in 1762 with green coats and red distinctives, which they wore until 1767. The Sacken Dragoons wore red coats with black distinctives and white metal buttons. In 1767 red coats were authorized for the Kurland Regiment and its cuffs were popinjay green plus and the coat had yellow metal buttons.

The Artillery, Pioneers and Transport Corps

In 1691 the Saxon artillery wore gray coats with red cuffs and collars, and red cloth stockings. Their hats had a cord. In 1717 they wore green coats with red collars, lapels, and cuffs, and straw-colored undergarments. The green and white uniform was worn until 1914, except for a short period from 1729 to 1730, when the field artillery had straw-colored distinctives. The uniform had yellow buttons with gold embroidery for the officers and NCOs and was in a style very much like that of the infantry.

The Electorate of Saxony at the End of the Thirty Years War

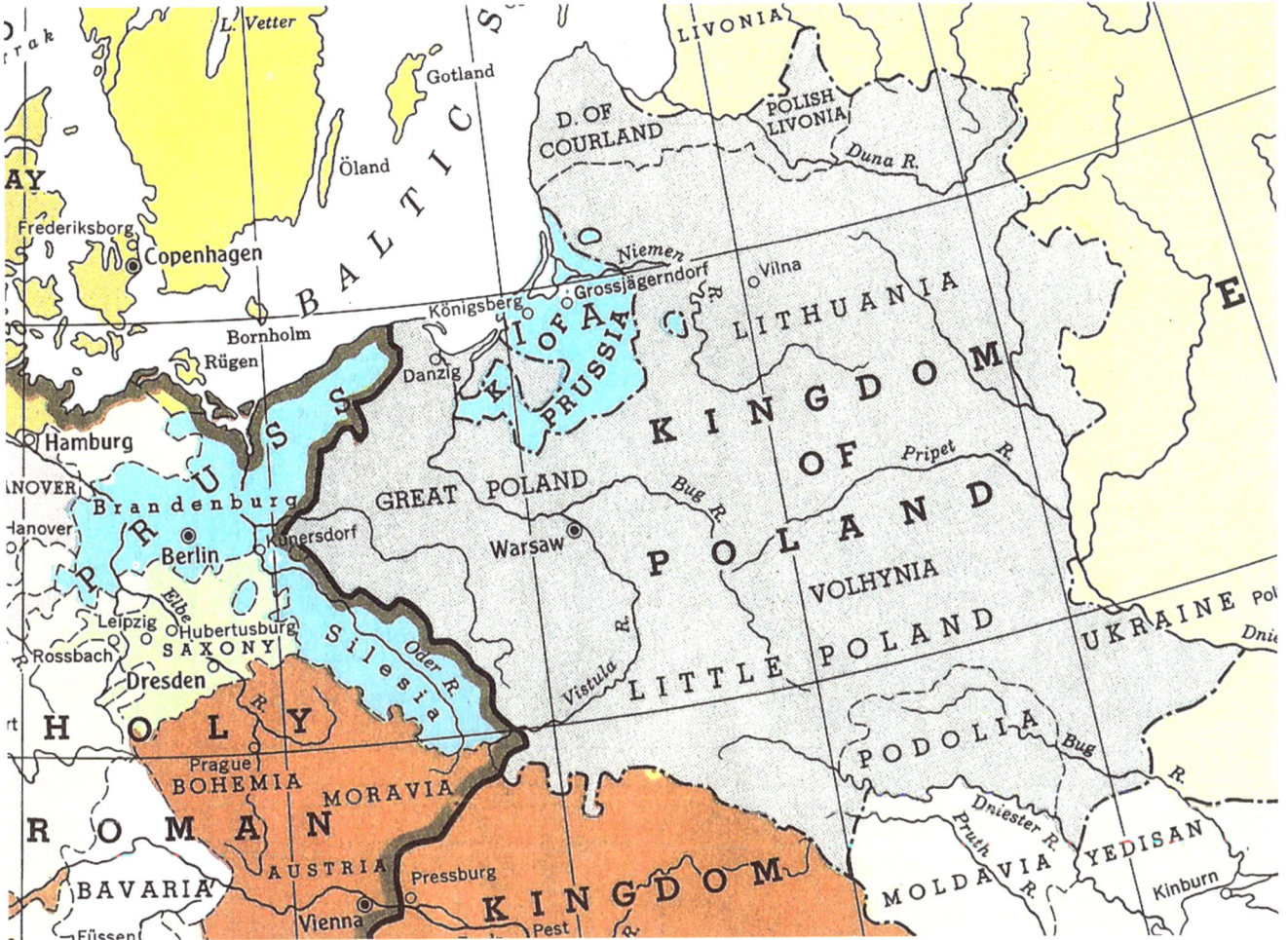

The Electorate of Saxony and The Polish - Lithuanian Commonwealth in 1763

1 – Leib Cuirassier Regiment in Gala
Uniform – 1699

2 – Leib Cuirassier Regiment Service
Dress – 1699

3 – Leibgarde zu Pferd Regiment –
Drummer – 1699

4 – Leibgarde zu Pferd – Officer – 1699

5 – Leibgarde zu Pferd – 1699

6 – Company of Giant Musketeers – 1700

7 – Swiss Leibgarde – 1700

8 – Officer from the 1st Guard Regiment – 1701

9 – NCO Musketeer from 1st Guard
Regiment – 1701

10 – Grenadier of the Saxon or 1st Guard
Regiment – 1701

11 – NCO from the 2nd Guard or von Bose Grenadier Regiment – 1701

12 – Grenadier from the Leib Regiment – 1701

13 – Field Infantry Regiment – 1705

14 – Wachtmeister from the Leib Dragoon
Regiment – Gala Uniform – 1706

15 – Dragoon Officer – 1706

16 – Dragoon – 1706

17 – Officer from the von der Goltz
Dragoon Regiment – 1707

18 – Officer from the von Damitz
Cuirassier Regiment – 1707

19 – Infantry Staff Officer – 1711

20 – Musketeer from the Graf Flemming
Infantry Regiment – 1711

21 – Soldier of the Land Militia – 1712

22 – Giant Musketeer – 1715

23 – Officer of the von Beichlingen
Infantry Regiment – 1715

24 – Officer of the von Friesen Musketeer
Regiment - 1717

25 – General – 1736-1753

26 – General Major – 1735-1753

28 – Leibgrenadier Garde – Fifer – 1745

27 – Brigade Major and Exercier Meister –
1735-1753

29 – Guardsman of the 1st Garde
Regiment – 1741

30 – Drummer of the 1st Garde Regiment – 1745

31 – Officer of the 2nd Garde Regiment – 1741   32 – Königin Infantry Regiment – 1740-1757

33 – Musketeer of the Sachsen-Weisenfels
    Infantry Regiment – 1745

34 – Graf Brühl Infantry Regiment – 1745

35 – Graf Stolberg Infantry Regiment –1748

36 – Garde du corps – Officer in Gala
Uniform – 1745

37 – Garde du Corps – Officer in Interum
Uniform – 1745

38 – Garde du Corps – Service Uniform – 1745

39 – Garde du corps – Exercise Collar – 1745

40 – Carabinier Garde – Trumpeter – 1744

41 – Carabinier Garde – Officer –1736-1754

42 – Carabinier Garde – 1735-1758

44 – Trumpeter of the von Plötz Cuirassier
Regiment – 1745

43 – Standarten Junker of the von Rechenberg
Cuirassier Regiment – 1745

45 – Oberst of the von Vitztnum Cuirassier
Regiment – 1746

46 – Drummer of the Chevalier de
Saxe Dragoon Regiment – 1745

47 – Officer of the von Rutowski
Chevauxleger Regiment – 1745

48 – Prinz Albert von Sachsen Teschen
Chevauxleger Regiment – 1745-1753

49 – Drummer of the Graf Brühl
Chevauxleger Regiment – 1745

50 – Artillery Officer – 1745

51 – Miner Corps – 1748

52 – Regimental Judge (Auditor) – 1745

53 – Leibgrenadier Garde – Officer – 1740

54 – Leibgrenadier Garde –Service Uniform
Ordinary Hat – 1740

55 – Grenadier of the Graf Cosal
Regiment – 1740

56 – Musician of the Grenadier Company of
the Grafen Bomnitz auf Sorau – 1740

57 – Fusilier of the von Rochow
Regiment – 1740

58 – Feldwebel of the de Caila
Regiment – 1740

59 – von Arnim Cuirassier Regiment – 1740

60 – NCO Curland Dragoon
Regiment – 1740

61 – Arnstädt Dragoon Regiment –
1735 - 1748

62 – Officer of the von Leipziger Dragoon
Regiment – 1740

63 – Staff Officer of the Graf Brühl
Chevauxleger Regiment – 1740

64 – General – 1756

65 – Leibgard Grenadier – 1750

66 – NCO of the 1st Kreisregiment Brutzen – 1748-1763

67 – Officer 2nd Kreisregiment Oschatz –
1748-1763

68 – 3rd Kreisegiment Weissenfels - 1758

69 – 4th Kreisregiment Freiberg – 1758

70 – Scharfschützen – Officer – 1775

71 – Scharfschützen – 1775

72 – Semi-Invalid Company
Waldheim – 1752

73 – Mounted Feldjäger – 1750

74 – Carabinier Regiment – NCO – 1758

75 – Farrier of the Leib Cuirassier
Regiment – 1756

76 – Von Schill Frei Hussar Squadron –
First French Uniform – 1760

77 – Von Schill Hussars –
German Uniform – 1763-1766

78 – Palace Swiss Guard –
Gala – 1764

79 – Palace Swiss Guard
(Daily Service) – 1764

80 – Noble Cadet Corps
(Parade Uniform) – 1764

81 – Noble Cadet Corps
(Daily Uniform) – 1764

82 – Fifer of the Leibgrenadier Garde
Regiment – 1764

83 – Officer of the Leibgrenadier Garde Regiment – 1764

84 –NCO of the Leibgrenadier Garde Regiment – 1764

85 – Grenadier of the Leibgrenadier Garde
Regiment – 1764

86 – Fifer of the Kurfürst Infantry
Regiment – 1764

87 – Officer of the Kurfürst Infantry
Regiment– 1764

88 – NCO of the Kurfürst Infantry
Regiment – 1764

89 – Grenadier of the Kurfürst Infantry
Regiment – 1764

90 – Musketeer of the Kurfürst Infantry
Regiment – 1764

91 – Trumpeter of the Garde du Corps
Regiment – 1764

92 – Officer of the Garde du Corps
Regiment – 1764

93 – Reiter of the Garde du Corps
Regiment – 1764

94 – Reiter of the Garde du Corps Regiment
(Interim Uniform) – 1764

95 – Reiter of the Carabinier Regiment – 1764

96 – Trumpeter of the Kurfürst Cuirassier Regiment – 1764

97 – Officer of the Kurfürst Cuirassier
Regiment – 1764

98 – Reiter of the Kurfürst Cuirassier
Regiment – 1764

99 – Musician of the Herzog Curland
Chevauxleger Regiment – 1764

100 –Officer of the Herzog Curland
Chevauxleger Regiment – 1764

101 – Reiter of the Herzog Curland
Chevauxleger Regiment – 1764

102 – Feldjäger – 1764

103 – Fifer of the Artillery Corps – 1764

104 – Officer of the Artillery Corps – 1764

105 – NCO of the Artillery Corps – 1764

106 – Gunner of the Artillery Corps – 1764

107 – Officer of the Engineer Corps – 1764

108 – NCO of the Engineer Corps – 1764

109 – Semi-Invalid Company – 1764

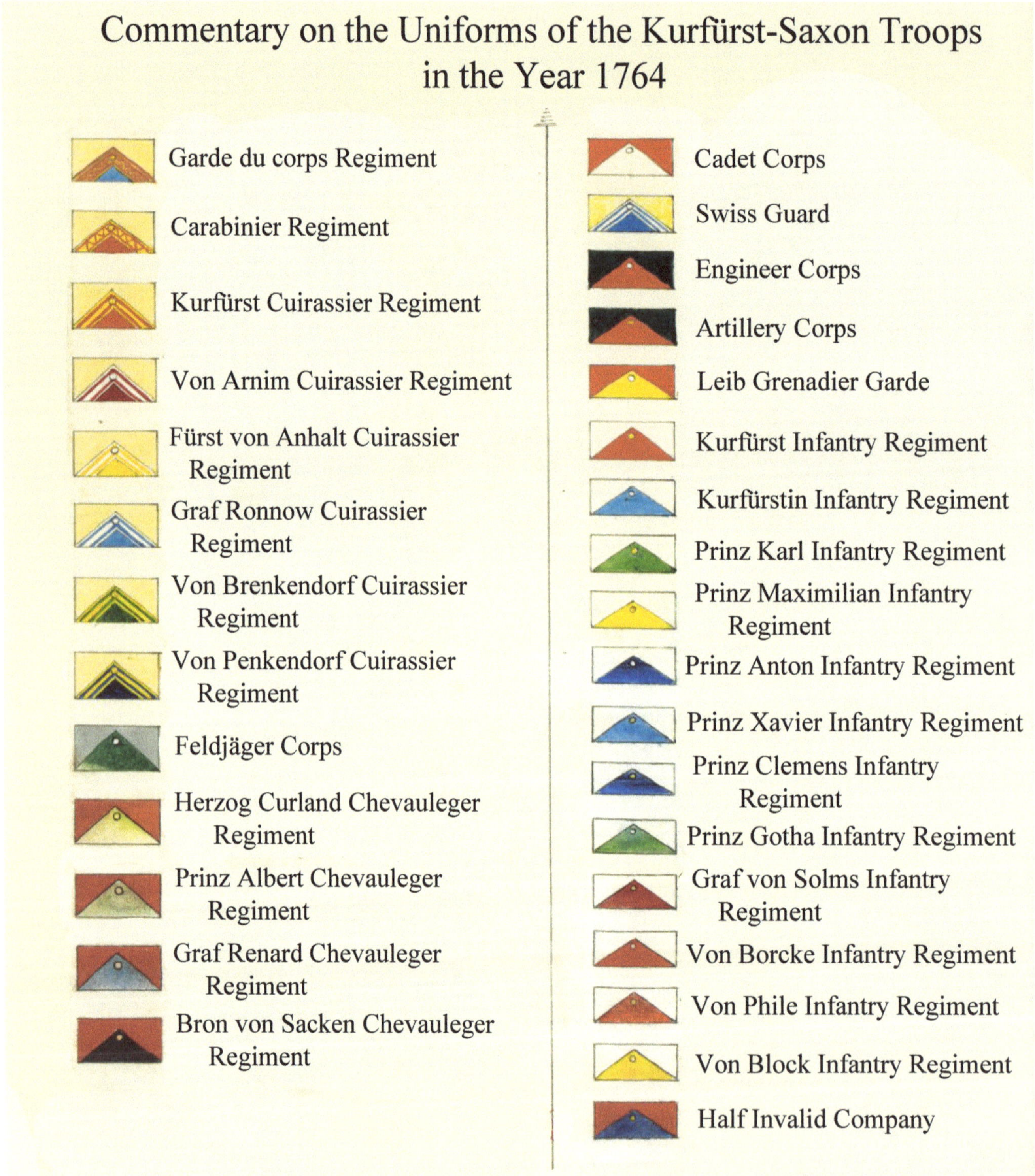

Commentary on the Uniforms of the Kurfürst-Saxon Troops in the Year 1764

Garde du corps Regiment

Carabinier Regiment

Kurfürst Cuirassier Regiment

Von Arnim Cuirassier Regiment

Fürst von Anhalt Cuirassier Regiment

Graf Ronnow Cuirassier Regiment

Von Brenkendorf Cuirassier Regiment

Von Penkendorf Cuirassier Regiment

Feldjäger Corps

Herzog Curland Chevauleger Regiment

Prinz Albert Chevauleger Regiment

Graf Renard Chevauleger Regiment

Bron von Sacken Chevauleger Regiment

Cadet Corps

Swiss Guard

Engineer Corps

Artillery Corps

Leib Grenadier Garde

Kurfürst Infantry Regiment

Kurfürstin Infantry Regiment

Prinz Karl Infantry Regiment

Prinz Maximilian Infantry Regiment

Prinz Anton Infantry Regiment

Prinz Xavier Infantry Regiment

Prinz Clemens Infantry Regiment

Prinz Gotha Infantry Regiment

Graf von Solms Infantry Regiment

Von Borcke Infantry Regiment

Von Phile Infantry Regiment

Von Block Infantry Regiment

Half Invalid Company

110 – Color Scheme for the Uniforms of the Kurfürst-Saxon Troops – 1764

77

111 – Swiss Leib Garde – 1735

112 – Lieutenant of the Prinz Xavier
Regiment – 1735

113 – Garde du corps – Officer – 1734 – 1738

114 – Garde du corps – Trabant – 1735-1738

115 – von Haxthausen Infantry
Regiment – 1736

116 – Saxon-Polish Uhlan – 1736

117 – von Sibilski Chevauxleger
Regiment - 1736

118 – Noble Cadet Company – 1765

119 – Noble Cadet in Interim Uniform – 1765   120 – Swiss Guard in Gala Uniform – 1765

121 – Swiss Guard in Daily Uniform – 1765

122 – Fifer of Leibgrenadier Garde – 1765

123 – Senior Officer of the Leibgrenadier
Garde – 1765

124 – Junior Officer of the Leibgrenadier
Garde – 1765

125 – Leibgrenadier Garde – 1765

126 – Fifer of Leibgrenadier Garde – 1765

127 – Senior Officer of the Leibgrenadier
Garde – 1765

128 – Junior Officer of the Leibgrenadier
Garde – 1765

129 – Grenadier of the Kurfürst Infantry
Regiment – 1765

130 – Junior Officer of the Kurfürstin Infantry
Regiment – 1765

135 – Prinz Clemens Infantry Regiment – 1765

136 – Senior Officer of the Prinz Gotha Infantry Regiment – 1765

137 – Graf Solms Infantry Regiment – 1765

138 – von Borcke Infantry Regiment – 1765

139 – Musketeer of the von Thiele Infantry
Regiment – 1765

140 – Von Blocke Infantry Regiment – 1765

141 – Garde du Corps – Trumpeter – 1765

142 – Senior Officer Garde du Corps – 1765

143 – Garde du corps – 1765

144 – Garde du corps in Interim Uniform – 1765

145 – Carabinier Regiment – 1765

146 – Trumpeter of the Kurfürst Cuirassier
Regiment – 1765

147 – Senior Officer of the Kurfürst Cuirassier
Regiment – 1765

148 – Kurfürst Cuirassier Regiment – 1765

149 – Von Armin Cuirassier Regiment – 1765

150 – Fürst Anhalt Cuirassier Regiment – 1765

151 – Graf Ronnow Cuirassier Regiment – 1765

152 – Von Penkendorf Cuirassier Regiment – 1765

153 – Von Brenkenhof Cuirassier
Regiment – 1765

154 – Feldjäger – 1765

155 – Musician of the Herzog Curland
Chevauxleger Regiment – 1765

156 – Senior Officer of the Herzog Curland
Chevauxleger Regiment – 1765

157 – Herzog Curland Chevauxleger
Regiment – 1765

158 – Prinz Albert Chevauxleger
Regiment – 1765

159 – Graf Renard Chevauxleger
Regiment – 1765

160 – Baron von Sacken Chevauxleger
Regiment – 1765

161 – Senior Officer Engineer Corps – 1765

162– Junior Officer Engineer Corps – 1765

163 – Fifer of the Artillery Corps – 1765

164 – Senior Officer of the Artillery Corps – 1765

165 – NCO of the Artillery Corps – 1765

166 – Gunner of the Artillery Corps – 1765

167 – Semi-Invalid Company – 1765

Colors and Standards of the Saxon Army

During the early part of the 18th century each company of infantry, dragoons and cuirassiers had a flag usually in the regimental distinction. After 1715 this was changed to one standard per squadron and two per battalion. The poles were usually painted black and the flags were affixed with brass nails.

In the period up to 1700 the standard was in the color of the unit distinction with "FA" embroidered in gold surrounded by a green wreath and gold flames in the corners.

After 1715 for battalions this was usually one with the unit distinction and one with a white background. The center was often a monogram of the Elector and on the reverse was the coat of arms in the Saxon-Polish-Lithuanian coat of arms. The central device was surrounded by green leaves. The corners and side panels often had flames in gold. The flag was square.

Cavalry standards were usually swallow tailed, but of similar design to infantry flags. They did not have flames but were usually heavily embroidered with a gold border

Examples of standards of the Saxon Army from, *Die Armee Augustus des Starken*. (Top) The infantry banner of the Infantry Regiment du Callia. (Bottom) banner of the Infantry Regiment Saxe-Weimer.

Examples of standards of the Saxon Army from, *Die Armee Augustus des Starken.*
(Top left) The standard of the Dragoon Regiment Chevalier de Saxe (top right) banner center device of Garde du Corps. (Bottom) The standard of the Garde du Corps

The State Flag of the Electorate of Saxony

Augustus II - House of Wettin (12 May 1670 – 1 February 1733), also known as Augustus the Strong, was Elector of Saxony from 1694 as well as King of Poland and Grand Duke of Lithuania in the years 1697–1706 and from 1709 until his death in 1733.

Augustus supposedly had great physical strength which earned him the nicknames "the Strong", "the Saxon Hercules" and "Iron-Hand". He liked to show that he lived up to his name by breaking horseshoes with his bare hands. He is also notable for fathering a very large number of children outside his marriage.

After the death of Jan Sobieski, he was elected king of the Polish–Lithuanian Common-wealth, and converted to Roman Catholicism. As a Catholic, he received the Order of the Golden Fleece from the Holy Roman Emperor and established the Order of the White Eagle, Poland's highest distinction. As Elector of Saxony, he is perhaps best remembered as a patron of the arts and architecture. He transformed the Saxon capital of Dresden into a major cultural centre, at-tracting artists from across Europe to his court. Augustus also amassed an impressive art collec-tion and built lavish baroque palaces in Dresden and Warsaw. In 1717 he served as the Imperial vicar of the Holy Roman Empire.

After poland was knocked out of the Great Northern War he was diposed from the throne and only regained the throne in 1709 with the defeat of Sweden, but led to stronger influnece of Russia in Poland-Lithuania. His tried to increase royal power in the Commonwealth, through broad decentralization as opposed to other European monarchies. He tried to accomplish this goal using foreign powers and thus destabilized the state. Augustus' death in 1733 sparked the *War of the Polish Succession.*

Augustus' body was buried in Poland's royal Wawel Cathedral in Kraków, but his heart rests in the Dresden Cathedral. His only legitimate son, Augustus III suceeded to the Electorate of Saxony and King of Poland-Lithuania in 1733. Several of his illegitimate sons went on to lead armies in the war of the 18th century, most prominently Maurice de Saxe.

Augustus III - House of Wettin (17 October 1696 – 5 October 1763) was the son of Augustus II of Saxony. From 1733 to 1763 he was the King of Poland and Grand Duke of Lithuania as well as Elector of Saxony in the Holy Roman Empire where he was known as Frederick Augustus II

He was the only legitimate son of Augustus II the Strong, and converted to Roman Catholicism in 1712 to secure his candidacy for the Polish throne. In 1719 he married Maria Josepha, daughter of Joseph I, Holy Roman Emperor, and became Elector of Saxony following his father's death in 1733. His desire to be ruler of the Polish-Lithuanian Commonwelath led to The War of Polish succession in 1733. He was able to gain the support of Charles VI by agreeing to the Pragmatic Sanction of 1713 and also gained recognition from Russian Empress Anna by supporting Russia's claim to the region of Courland. He was elected king of Poland by a small minority on 5 October 1733 and subsequently removed former Stanisław I Leszczyński who was also elected king during the Great Northern War. He was crowned in Kraków on 17 January 1734.

Augustus was supportive of Austria against Prussia in the War of the Austrian Succession (1742) and again in the Seven Years' War (1756), both of which resulted in Saxony being defeated and occupied by Prussia. In Poland, his rule was marked by the increasing influence of the Czartoryski and Poniatowski families, and by the intervention of Catherine the Great in Polish affairs. His rule deepened the social anarchy in Poland and increased the country's dependence on its neighbours, notably Prussia, Austria, and Russia. Upon his death Catherine the Great prevented his son from being elected to the Polish throne, supporting her former lover, Stanisław August Poniatowski, instead.

Throughout his reign, Augustus was known to be more interested in ease and pleasure than in the affairs of state; this notable patron of the arts left the administration of Saxony and Poland to his chief adviser, Heinrich von Brühl, who in turn left Polish administration chiefly to the powerful Czartoryski family.

Other books from the Nafzinger Collection

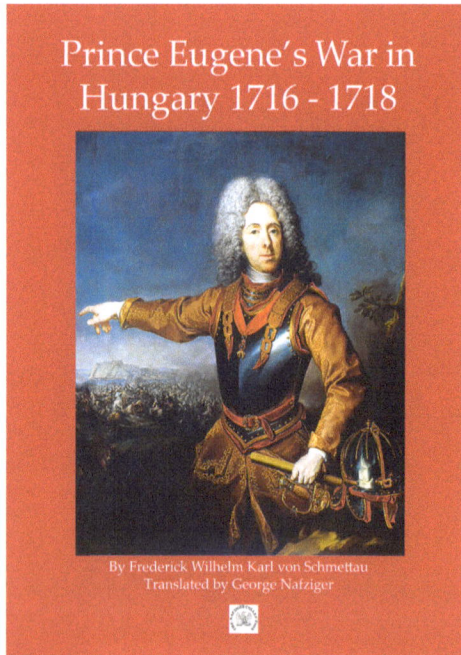

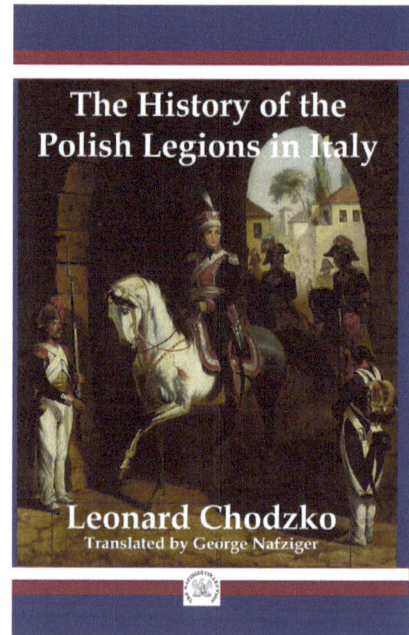

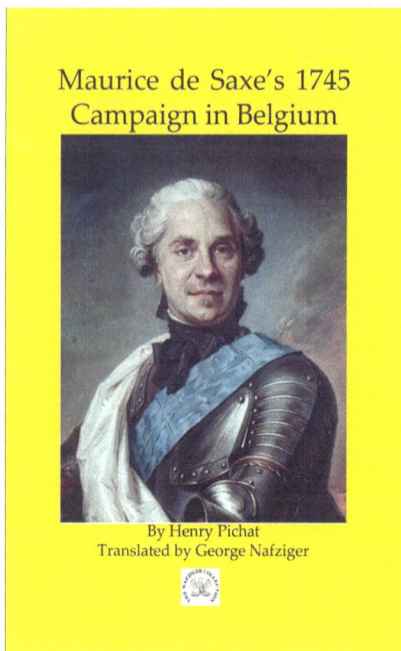

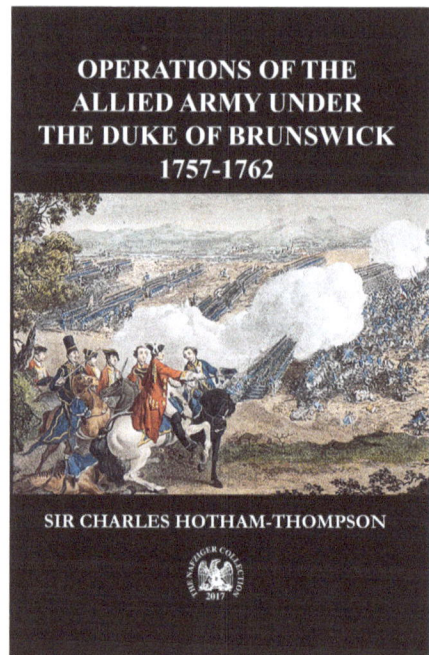

Prince Eugene's War in
Hungary 1716 - 1718

By Frederick Wilhelm Karl von Schmettau
Translated by George Nafziger

The History of the
Polish Legions in Italy

Leonard Chodzko
Translated by George Nafziger

Maurice de Saxe's 1745
Campaign in Belgium

By Henry Pichat
Translated by George Nafziger

OPERATIONS OF THE
ALLIED ARMY UNDER
THE DUKE OF BRUNSWICK
1757-1762

SIR CHARLES HOTHAM-THOMPSON

Look for more books from Winged Hussar Publishing, LLC – E-books, paperbacks and Limited-Edition hardcovers. The best in history, science fiction and fantasy at:

https://www. wingedhussarpublishing.com
https://www.whpsupplyroom.com

or follow us on Facebook / Instagram at:

Winged Hussar Publishing LLC

Or on twitter at:

WingHusPubLLC

For information and upcoming publications

www.ingramcontent.com/pod-product-compliance
Lightning Source LLC
Chambersburg PA
CBHW041955100426
42812CB00018B/2655